The Complete Guide to

SCIENTIFIC MANUSCRIPT WRITING

Andrea R. Gwosdow, Ph.D.

AVIVA
PUBLISHING
New York

The Complete Guide to Scientific Manuscript Writing
© 2018 Andrea R. Gwosdow, Ph.D.

Published by:
Aviva Publishing
Lake Placid, NY
(518) 523-1320

www.AvivaPubs.com

Print ISBN: 978-1-947937-32-1
Library of Congress Control Number: 2018913573

Cover and book design: Meredith Lindsay/MediaMercantile.com

Contents

Introduction

I have always been interested in science. I was drawn to laboratory science because I liked the scientific process of making an hypothesis, figuring out how to test the hypothesis, setting up and running experiments, and analyzing the data to ultimately discover a new piece of information. It was the thrill of learning new information that no one else in the world knew that was my favorite part of the scientific process. That's why I went into scientific research–to make discoveries that would help improve our lives.

My research was initially aimed at understanding how we adapt to different environments, such as heat or cold. The process of adaptation involves the stress response and my research focused on the interaction between stress hormones and inflammatory agents from the immune system. As a scientist, I understood early in my career that I could not just do experiments and learn new information without communicating my work.

Communication is a huge part of our work. As scientists, we need to communicate and share our work with other scientists and clinicians as well as with the general public. This allows us to discuss our work with colleagues and educate others while we disseminate our work.

I have always loved writing. As a scientist I was always the person in the laboratory who published data promptly. Often people asked me for help writing their manuscripts. I found I enjoyed teaching the writing process.

Writing is a major avenue for sharing scientific work. Typically, new information is first shared at a scientific or clinical meeting. Each meeting has a "call for abstracts" and abstracts are written for submission. Once the abstract is accepted for presentation, a poster or oral (slide) presentation is prepared for the meeting. This is the first public sharing of new scientific and clinical data.

The next step is to prepare the data for publication in a scientific or clinical journal. Writing a manuscript can be a complex undertaking. The process of scientific and clinical writing requires focused time and energy to tell the scientific story of the new discovery. This involves accurately presenting the data and explaining the importance of the findings within the context of the current scientific or clinical literature.

Many scientists and clinicians do not enjoy writing. Graduate students, residents, and fellows typically learn writing from an advisor, mentor, or preceptor. Often these people are very busy and have varying amounts of time to devote to teaching you how to write scientific documents. As a result, the process of writing may be frustrating and arduous. A clear explanation of what to write in each section of the manuscript would make your job easier and more enjoyable.

This book is that explanation. It began when I was invited to teach scientific and clinical writing to graduate students and residents at a university in Portugal. I read many books on the subject. All were technical and overly detailed. As a seasoned scientist and professional medical writer, I did not find these books helpful.

I decided to focus on my own writing process and develop my own writing course. I organized my writing process into a series of writing guidelines. This book contains the guidelines I use every day to communicate my work and those of my clients when I write scientific and clinical manuscripts. They include guidelines and recommendations for writing each section of the manuscript along with templates for the outline and manuscript. My workshops are interactive because I believe people learn best by doing and practicing, which makes writing easier and more fun. In order to practice, I have developed activities for workshop participants using these guidelines.

This book grew out of my manuscript writing workshops which I have presented internationally. This book represents over thirty years of experience writing more than sixty manuscripts for publication in peer-reviewed scientific and clinical literature. In this book I offer general writing guidelines to use throughout your writing process as well as a format for writing each section of the manuscript. I have also adapted activities from my workshops for you to practice using the guidelines presented in each chapter.

These guidelines have helped many graduate students, scientists, and clinicians learn how to write journal articles. If you are already familiar with this process, my guidelines and recommendations will help you write more efficiently and effectively. These guidelines can improve your writing and make it easier for you to communicate your work.

These writing guidelines can also be used to evaluate the scientific literature. Learning how to evaluate the scientific literature will improve your writing. It will also help you decide what papers to include in your literature review, introduction, and/or discussion sections.

You can read the book cover to cover or, if you prefer, you can use it to write each section of the manuscript. By following the guidelines in this book, a completed first draft of your manuscript can be written.

Please note this book is copyrighted. You may use it to write your own journal articles. However, reproduction and distribution of this book for other reasons is prohibited except with my written permission. Please contact me at andrea@gwosdow.com to discuss permissions.

For information on my writing workshops and services, please visit my website www.gwosdow.com. Sign up for my newsletter to receive regular writing guidelines and tips, suggest topics for future newsletters, and share your writing tips.

Good luck writing your manuscripts. Please send feedback or suggestions to me at andrea@gwosdow.com.

About the Author

Dr. Andrea Gwosdow has a long history of helping researchers communicate science. In 1997, Dr. Gwosdow founded Gwosdow Associates Science Consultants LLC, an award-winning science and medical communications company. Gwosdow Associates is dedicated to providing medical and science writing and training services, working with schools on science education, and interpreting science for non-scientists.

Dr. Gwosdow is an effective medical writer who has received the top awards from the National and New England Chapter of the American Medical Writers Association (AMWA) and 2nd place in the 2014 R&D Pharmaceutical Awards for Clinical Writing. Dr. Gwosdow is an active member of both AMWA and the American Physiological Society (APS). She has served on committees in both organizations and has been leading workshops at AMWA's annual conference since 2004.

Dr. Gwosdow is known for her clear explanations of complex scientific principles to a variety of audiences at all levels. Dr. Gwosdow has worked around the world with a variety of organizations to improve their communication and writing skills through hands-on workshops, seminars, and presentations. Dr. Gwosdow also writes a monthly column about cutting edge research for middle and high school students and teachers (www.whatayear.org), and organizes APS's Physiology Understanding (PhUn) Day at Boston Children's Museum.

Dr. Gwosdow is a skilled scientist with experience leading research teams in systems physiology, endocrinology, and immunology. She has expertise in animal, cellular, and molecular biology as well as data collection and analysis of laboratory samples. Dr. Gwosdow maintains her academic appointment as a Lecturer at Harvard Medical School and a Senior Scientist at Massachusetts General Hospital through her involvement in medical education. Dr. Gwosdow has developed and managed an award-winning business/ school partnership and has published over twenty-two articles on her research in peer-reviewed journals such as the *American Journal of Physiology, Endocrinology, Laboratory Animal Science, Advances in Physiology Education, and Science Scope.*

Dr. Gwosdow's newsletter and blog, Writing Tips and More, offers suggestions and tips on science communication to make the details and intricacies of communicating science available to everyone interested in this topic.

Framework of this Book

This book is divided into 16 chapters. Here is a view of each chapter:

CHAPTER	TITLE	FOCUS
1	General Writing Guidelines	Know general guidelines before you begin writing your journal article.
2	Anatomy of a Journal Article	Read an overview of each section of a journal article with guidelines for organizing and writing more effective journal articles.
3	Reading the Scientific Literature	Understand how a journal article is read to help you improve your writing process.
4	Outlines	Learn the importance of outlining and its role in writing journal articles.

CHAPTER	TITLE	FOCUS
5	Writing the Introduction	Learn practical recommendations for writing the introduction with activities for practice.
6	Writing the Materials and Methods	Learn the basics of writing the materials and methods with activities for practice.
7	Writing the Results	Learn how to write the text and present your results clearly as you practice making, analyzing, and interpreting tables and figures.
8	Writing the Discussion and Conclusion	Practice writing concise and clear discussion paragraphs using the guidelines for writing the discussion and conclusion.
9	Writing the Abstract	Learn the basics of writing formatted abstracts.
10	Selecting a Title	Learn the importance of the title and how to select one.
11	Citing References	Learn to distinguish the types of references and when to cite them.
12	Now that the Manuscript is Written–Revise	Learn a process for revising your journal article.

CHAPTER	TITLE	FOCUS
13	Authorship: Getting Credit for Your Work	Understand the guidelines for authorship.
14	Responding to Reviewers' Comments	Understand reviewers' comments and learn a process for responding to them.
15	Writing the Poster	Learn the basics of writing a poster.
16	Presenting Your Poster and Yourself	Learn the basics of presenting your poster and yourself.
Appendix 1	All Writing Guidelines	A summary of all writing guidelines presented.
Appendix 2	Outline and Manuscript Template	A template to follow when writing your outline and journal article.

Chapter 1:
General Writing Guidelines

LEARNING OBJECTIVES

- Identify five guidelines for writing.

- Identify a process to incorporate writing into your regular work schedule.

Chapter Highlights:
GUIDELINES FOR GENERAL WRITING

- Identify your audience and target journal before you begin writing.
- Write for the readers of your target journal.
- Start by writing the section that is easiest to write.
- Schedule time to work on your writing each day.
- Write one section of the manuscript each day.
- At the end of one writing session, do not look at your writing for at least one day.
- Re-read, edit, and revise what you have written.
- Read your revision out loud to hear how it flows and sounds.
- Repeat this process until you are happy with your writing.
- When revisions are complete, give the manuscript to a colleague or peer to read, review, and provide feedback.
- Repeat process until manuscript is completed.
- Start with an outline.
- Use simple short sentences: subject-verb-object.
- Use complete sentences and paragraphs.
- All paragraphs flow in a logical order.
- Connect sentences and paragraphs with transition words.
- Start a new sentence by repeating words and phrases from the previous sentence.
- Define all abbreviations. Use a maximum of four.
- Check all spelling, punctuation, grammar, word choice, capitalization, and word order.

Before you start writing, here are some general writing recommendations:

Identify your audience and target journal before you begin writing. Write for the readers of your target journal.

Before you begin writing your manuscript,[1] identify your audience. This will determine the language you use in your writing. For example, if you are writing your manuscript for a specific journal, using technical terms may be acceptable. However, if the manuscript is aimed at a more general audience, you will need to identify your terms and include additional information so a general audience can understand your article. You can identify your audience and target journal by asking the following questions:

- Who would you like to read this manuscript–scientists, clinicians, the public, others?

- Who reads your target journal?

When selecting a target journal, make sure the readers of this journal are the people you want to read your manuscript. Align the readership of the manuscript with your target journal *before* you begin writing.

As you start writing for your target journal, pay careful attention to the format required by the journal. Write in the format required by your target journal. This includes adhering to any word limitations (total word count or word count by section), numbers of tables, figures or references, special sections such as highlights, keywords, acknowledgements, or disclosures. The target journal's expectations are found in the guidelines for authors.

---

[1] A manuscript is a paper that has not been published (American Medical Association Manual of Style, 2007).

Start by writing the section that is easiest to write.

The section that is easiest to write may be different for each person. For example, if this is the materials and methods section, this part can be written while you are finishing your experiments. If this is the results section, you can begin to put the tables and figures together once the data have been collected. For the results section, many people begin by drawing graphs and tables and deciding what data will be included in the manuscript. Starting the manuscript by writing the section that is easiest for you is an important step. As long as the information is placed in the desired section, the paper can be written in any order.

Schedule time to work on your writing each day.

Many people feel they need a lot of time to write their manuscript. For this reason, they have long writing sessions after their experiments are completed. While this may work for some people, others are more productive if they schedule small amounts of time (one to two hours) to work on their writing several days a week. Book your writing into your daily schedule and keep this "date" with yourself. This way, you will have the time set aside to work on the project. Getting into a working rhythm is an important step in writing a manuscript.

Write one section of the manuscript each day.

Separate your writing into small sections that are doable in a short amount of time. One or two hours is manageable. If you do this each day, by the end of the week you will have worked on your project for many hours, and you will see progress. I always feel a sense of accomplishment when I do this, and I hope you will too.

At the end of each writing session, write a few sentences of the next section to put your ideas on paper. This helps prevent writing inertia by providing reminders of your thought process when you return for the next writing session.

At the end of one writing session, do not look at your writing for at least one day.

Schedule your writing, write for the selected amount of time or until you finish a section, then stop writing and let your writing "incubate" for at least a day. This gives you time to focus on other things and gets your mind off your writing. When you remove yourself from your initial writing, you can let it settle, and when you come back to your writing, you will see it in a different light. You will then be able to evaluate your writing more objectively.

Re-read, edit, and revise what you have written.

After the "incubation" period, reread your writing from your last writing session. Then, read it aloud, so you can hear how it sounds. Hearing your words helps your writing. Many times, after hearing my writing, my "pearls of wisdom" from the day before are ready to be rewritten and revised.

Revise your writing by asking yourself some questions:

- Did I understand what I wrote? If I didn't, that tells me what I must do next.

- Is that what I meant to say?

- Could I say it better using different words?

I find that I am a harsh critic which enables me to catch many of my own errors before someone else reads it.

Repeat this process until you are happy with your writing.

Follow the steps just discussed until you have a draft manuscript you are ready to show to others. Let's review these steps:

- Schedule time to write.

- Write one section of a paper each day.

- Let your words incubate for one day.

- The next day, reread your work and revise what was written.

- Repeat this process until you have a draft with which you are happy.

I usually edit my work at least three times before I am comfortable with it.

When revisions are complete, give the manuscript to a colleague or peer to read, review, and provide feedback.

This will enable you to get honest feedback from a colleague. These comments will be your first peer review comments. Take these comments seriously as you continue to revise your manuscript. Use this process until your manuscript is ready for submission to the target journal. Remember, the goal is to submit a manuscript with few errors that can be accepted with minimal revisions.

It is common for manuscripts to be returned by the journal with reviewers' comments. One goal of the reviewers' comments is to improve the manuscript. However, the better the manuscript is when submitted, the fewer reviewers' comments you may receive and the more likely your manuscript will be accepted for publication. By using these guidelines, you will improve your writing and produce better manuscripts for journal submission.

Chapter 2:
Anatomy of a Journal Article

LEARNING OBJECTIVES

- Describe how a journal article appears in a journal.

- Identify the main point of each section of a journal article.

Chapter Highlights:
GUIDELINES FOR A WELL-WRITTEN JOURNAL ARTICLE

- Aims are well-defined.
- Research question(s) and their importance are articulated.
- Significance of paper is explained.
- Hypothesis is clearly presented.
- Methods are short, concise, and understandable.
- Study design is easy to understand.
- Study endpoints are well-defined.
- Subheadings are used in materials and methods and results sections.
- Results are presented clearly.
- Results are interpreted accurately.
- Discussion is informative and focused on topic.
- Paper is well-referenced.
- Ideas flow in a logical sequence.
- Common abbreviations are used and defined.

Let's look at a journal article.[1] The parts of a paper as they appear in the journal are, in order: Abstract, Introduction, Materials and Methods,[2] Results,[3] Discussion,[4] and Conclusion. Familiarize yourself with the style of journals in your field by reading articles from your target journal before you start writing your manuscript.

---

[1] A journal article is a published paper (American Medical Association Manual of Style, 2007).

[2] Some journals place the materials and methods section after the conclusion. One example is *Nature Medicine*.

[3] Some journals combine the results and discussion sections.

[4] Some journals combine the discussion and conclusion sections.

What information is found in each section of the journal article? The summary below and Table 1 can serve as a guide for you to follow:

Abstract

The abstract summarizes the paper. It provides a short concise explanation of the study, including the findings and their significance. The first sentence states the objective of the paper. The last sentence indicates the conclusion of the study. The remaining sentences briefly explain the materials and methods and present the significant results. The exact length of the abstract may vary from journal to journal. An average length ranges between 250 and 350 words.

Introduction

The introduction introduces the general topic being studied, indicates what the unmet need is in this area of research, and presents the question(s) under investigation. The first sentence of this section introduces the topic. The last sentence presents the question or focus of this study. The middle sentences and paragraphs indicate the information missing in this area of research and state why the research is important.

Materials and Methods

The materials and methods section details how the study was conducted. This section provides the detailed information needed to understand the study design. It also includes enough information so other researchers can evaluate and replicate the study.

Results

The results section presents the findings in a logical organized manner. In a well-written journal article, the reader should feel as if the author is taking the reader by the hand and leading the reader through the tables and figures. The tables and figures are presented in a logical order that tells a story. This may not be the order in which the experiments were conducted. It is, however, the logical order in which the story is easiest to understand.

Discussion

The discussion section explains each finding and places each finding in the context of the known work conducted in this field. Each finding or result is discussed in one or more paragraphs in the same order it is presented in the results section. The discussion section includes a paragraph explaining the limitations of the study. It also links the study to the current and known literature, indicating the strengths and weaknesses of the knowledge in the field.

Conclusion

The conclusion provides the take home message of the paper. This is the sound bite and includes the main findings of the paper and their significance. It tells the reader why the study was important and the most significant finding(s). The conclusion emphasizes the value of the research and explains why the reader should care about the study. The conclusion is usually the last paragraph of the discussion section.

Each of these sections will be discussed in more detail in this book. Before we discuss writing, we first need to understand how a journal article is read.

Table 1. Anatomy of a Journal Article

Abstract: Summarizes the paper

Introduction: Presents the question(s) to be answered/ investigated and hypothesis

Materials and Methods: Details the methods used to determine the answer

Results: Presents the findings

Discussion: Considers the findings in the context of other work done in this field

Conclusion: Provides the take home message

Chapter 3:
Reading the Scientific Literature

LEARNING OBJECTIVES

- Identify the importance of reading journal articles.

- Identify one process for reading journal articles.

- List the order in which you read a journal article.

- List the order in which you write a journal article.

> ## Chapter Highlights:
> ### GUIDELINES FOR READING THE SCIENTIFIC LITERATURE
>
> - Read, read, and read lots of journal articles in your area to see how they are written.
> - Reading process:
>   - ‣ Read the abstract.
>   - ‣ Skim the paper.
>   - ‣ Read the journal article for details.
>   - ‣ Underline every word and phrase you do not understand.
>   - ‣ Look up words you do not understand.
>   - ‣ Reread the journal article for comprehension.
>   - ‣ Read the journal article out loud, so you can hear how it sounds and flows.

Read, read, and read lots of journal articles in your area to see how they are written.

Good writing begins with reading. Reading the scientific literature is as important as writing. Reading will familiarize you with the style and content of each journal. You will see how the material is presented in each section of the journal article, how the references are cited, and how the language is used. All these factors will help you with your own writing.

It is especially important to read recent journal articles from journals in your field. Pay special attention to the particular journal(s) you plan to use for your manuscript submission. This will allow you to familiarize yourself with the journal style and writing *before* you begin writing your own manuscript. This can save many hours of writing and rewriting before your manuscript is ready for submission to the journal.

Guidelines for Reading Journal Articles

People often ask, how do I start reading journal articles–they seem so long and complicated? Here are some guidelines and a process to get started.

- **Read the abstract.**
  Reading the abstract will give you the main point(s) in the paper without all the details. This will help you with the general overall picture of the message in the journal article.

- **Skim the paper.**
  Skimming the paper will give you the main ideas without getting lost in all the details. This will help you with the general overall picture and the message the authors are conveying in the journal article.

- **Read the journal article for details.**
  Once you have the general idea of the paper, go back and read every word and line in the journal article. This time you are reading the journal article for comprehension. Ask yourself the following questions:

  ▸ What is the hypothesis?

  ▸ Were appropriate materials and methods used?

  ▸ How was the study designed?

  ▸ Was the study designed appropriately?

  ▸ What results are presented?

  ▸ What are the main result(s) of the study?

  ▸ What evidence is presented to support this claim(s)?

  ▸ Would you like additional evidence to support this claim? If so, what?

> ‣ What are the limitations of the study?

> ‣ Is the take home message of the study accurate?

- **Underline every word and phrase you do not understand.**

  This can be done as you read the journal article. It will help you follow the thought process and story of the paper.

- **Look up words you do not understand.**

- **Reread the journal article for comprehension.**

  The journal article should be understood more clearly after this reading.

- **Read the journal article out loud, so you can hear how it sounds and flows.**

  A paper read aloud sounds different than read silently. This will train your ear to identify the style of the journal. Knowing the journal's style will guide you as you write your own manuscript.

Activity 1:
How do You Read a Journal Article?

Sections of a journal article:

Abstract	Discussion
Results	Materials and Methods
Introduction	Conclusions

1. Take any journal article.

2. Read this article as you usually read a journal article.

3. Notice the order in which you read the journal article. Note what section you read first, second, third, fourth, fifth, and sixth.

4. Put the parts of the journal article in the order you read them in the space below:

    •

    •

    •

    •

    •

    •

Note the order you used to read the journal article. Is this different from how the journal article is printed in the journal? If you answered "yes", you are not alone. Many people respond "yes" to this question. Many people do not read the journal article in the order presented by the journal. This is important to know because it will help you understand the rationale for including key information in particular places in your own manuscript.

Activity 2:
How do You Write a Journal Article?

Sections of a journal article:

Abstract	Discussion
Results	Materials and Methods
Introduction	Conclusions

1. Take any journal article.

2. Identify how you write a manuscript.

3. Notice the order in which you write a manuscript. In other words, what section do you write first, second, third, fourth, fifth, and sixth?

4. Put the parts of the manuscript in the order you write them in the space below:

-
-
-
-
-
-

Do you write a manuscript in a different order than it appears in the journal? If you answered "yes", you are not alone. Many people write manuscripts in a different order than they appear in the journal.

What does this mean? How a journal article appears in a journal is different from how it is read which is also different from how it is written. These are two reasons why writing manuscripts are challenging.

We need to keep this in mind as we examine the anatomy of the journal article and identify the components of each section of the manuscript. To be understood, information needs to be included in specific areas of the manuscript to account for the different ways journal articles are read.

Chapter 4:
Outlines

LEARNING OBJECTIVES

- Identify the importance of outlines.

- Identify a process for developing an outline.

- Identify a process for reviewing and revising an outline.

- Identify two recommendations for outlining.

- Use a simple outline to organize each section of the manuscript.
- Write each idea as a short bullet point.
- Just write the idea. Don't worry about the English.
- If the ideas have citations, include the references.

Use a simple outline to organize each section of the manuscript.

Often people get stuck trying to write the manuscript with perfect sentences and do not take the time to evaluate and discuss the results and the best way to present them. The result may be a frustrating manuscript writing process. However, if an outline is completed first, much time and energy is saved. I find simple outlines are critical in manuscript writing.

What is meant by outlining?

By outlining, I mean a bullet pointed list of the main thoughts and ideas to be included in each section of the manuscript. It does not have to be a formal outline. Don't get stuck on the actual outline; a bullet pointed list works fine. Just write the idea. Do not worry about the English because you can fix the grammar later. If the ideas you write have citations, include the references. This will help you later when you complete the references section. A template for writing an outline is provided in Appendix 2.

Why write an outline?

The outline helps you get your thoughts and ideas on paper. Think of each manuscript as telling a story. The flow of the story is very important. The outline is a good tool to use to evaluate and discuss the flow of the story with yourself and others.

The outline enables you to visualize the entire story and helps you decide if additional information is needed or if too much information has been included. The result is that outlining will save you time because you will take the time to plan the manuscript and carefully decide what to include in each section. The outline allows you to look at the manuscript as a whole and see if it accurately conveys your results and messages. If it does, you are ready to start writing! If it does not, move the information around before you write. The resulting document will be more focused and will require less work to produce the finished product.

The advantage of outlining is that it will help prevent common errors typically found in submitted manuscripts such as including results in the methods section, writing a literature review as the introduction section, or omitting pertinent pieces of information.

If you are writing with other people, outlining is a good way for all authors to add their ideas and evaluate the flow of the story. It also provides a document for all authors to discuss and revise. Once the outline is agreed upon by all team members, the writing can begin.

How do I start my outline?

I start outlining online using a Word file. Each section of the manuscript is on a separate page. This process works while I gather my thoughts and write the first draft of the outline. Once I have an outline I want to critique or share with others, I use the index card method described below. Post-it notes also work well.

Nuts and Bolts of Outlining

1. Select 8 different colors of index cards.

2. Label each colored index card with the section of a manuscript: Abstract, Introduction, Materials and Methods, Results, Discussion, Conclusion, References, Acknowledgements.

3. Put one sentence from the outline on the appropriate colored card.

4. Order the cards as you plan to write the manuscript.

5. Place the cards in this order.

6. Critique the order.

    a. Are you conveying your intended message?

    b. Does the order flow in a logical fashion?

    c. Do all the authors agree with this order?

Once an order is agreed upon, put the index cards down for at least a day. When you come back to the project the next day, you will look at it from a new perspective. Start by rereading what was written and/or agreed upon. In fact, read it aloud, so you can hear how it sounds. Hearing your words will help your writing.

As you review the cards for each section of the manuscript, ask yourself the following questions:

- Does the manuscript tell a story?
- Does the manuscript flow in a logical order?
- Does the manuscript convey the intended message(s)?
- Does the introduction explain why the study was conducted?
- Does the materials and methods section provide enough details for replication?
- Are the results presented in a logical order?
- Do the tables and figures clearly support the story?
- Do the results support the conclusions?
- Does the discussion indicate the significance of the study?
- Does the discussion place the results in the context of the scientific literature?
- Are the limitations of the study explained in the discussion section?

- Does the conclusion accurately summarize the study?
- Does the conclusion provide the desired take home message of the manuscript?

If you can answer all these questions positively, you are ready to get honest feedback from one or more colleagues. If not, take the time to revise your work. This will save you time later. Repeat the process until you and your colleagues are happy with your results.

This method works well because the index cards or post-it notes can be moved around easily. Hence, you can "try out" thoughts in different sections of the manuscript to see where they flow best.

Gaining Consensus

A draft outline can be used as the focus of an author team meeting. In this case, the goal of the team meeting is to have all the authors think about the results and messages of the manuscript. As a team, answer the questions on the previous page to ensure the manuscript is complete. This process can be repeated until all authors agree on the order of presentation of the results, the messages of the paper, and the flow of the story. Gaining consensus may take one or more meetings. Don't begin writing until consensus or an agreement has been reached. This will save you time and effort writing your manuscript.

Activity 3:
Outlining

1. Read one section of a published journal article.

2. On a piece of paper, colored post-it note, or index card write one main point and two to four supporting smaller points.

3. Repeat this process with other sections of the journal article. When you finish, read the main point of each section out loud. Does the main point of each section tell the story of the journal article?

4. Note the changes, if any, that you might suggest to improve the flow of the journal article. Being aware of these changes will help you improve your own writing.

Chapter 5:
Writing the Introduction

Chapter Highlights:
GUIDELINES FOR WRITING INTRODUCTIONS

- Keep introduction short: three paragraphs (nine sentences minimum).

- Explain the disease or area of research and introduce the topic briefly in the first paragraph.

- Start the first sentence of the first paragraph with a topic sentence.

- Introduce the area of research that is not known and explain the significance of this missing piece of information in the second paragraph.

- Clearly state the gap in knowledge, which is the topic of your manuscript.

- State the hypothesis in the third paragraph.

- Include a statement of the purpose and significance of the manuscript.

- All three paragraphs flow in a logical order that is easy to understand and follow.

- Reference all facts.

- Cite only literature relevant to your topic.

- Define all questionable terms and abbreviations.

- Do not mention results in the introduction.

The introduction is the first part of the journal article the reader sees after the abstract. It provides information to compel the reader to read the journal article. The goal of the introduction is to introduce the topic, explain what is known and not known in the specific area of research, why this information is important, what question(s) will be answered, and why the study was necessary. This is a lot to accomplish in a few paragraphs while you capture the attention of the reader! This chapter will give you the structure and tools to make this job easier.

Think of the introduction as a funnel consisting of three paragraphs with a minimum of three sentences in each paragraph. In reality, the introduction may be more than three paragraphs in length, but this is a good goal to work toward. As you will see below, the introduction has a specific purpose and needs to be focused to accomplish this goal.

The first paragraph introduces the topic or disease state and emphasizes its importance by providing a general concise description of the topic or disease state. Sometimes more than one paragraph is needed to describe the topic adequately, but this is only meant as a brief introduction to the topic. To introduce the topic, ask yourself the following question: what essential pieces of information does the reader need to know to understand the topic? This is the information that should be included in the first paragraph.

Start out the first sentence of the first paragraph with a topic sentence that introduces the subject. Some examples of sentence starters for topic sentences are below:

This disease is a genetic disease caused by a mutation in …
Peptidoglycan biosynthesis is…
Glucosylceramide is the precursor of …
Saposins are a group of proteins…

The second paragraph introduces the gap in knowledge that your study fills. Use the literature to explain this "hole" and the significance and importance of filling this gap. This paragraph answers the following questions: 1) Does it matter that this hole is in the literature, and 2) Why do we need to know this information? Sometimes more than one paragraph is needed to answer these questions and put this information into perspective. Be careful to include only the information that is needed for the reader to understand the gap in knowledge and why your experiments are important. Include only "must have" information in this paragraph. Other "nice to know" information can be included in the discussion section.

Start out the first sentence of the second paragraph with a sentence like one of the examples below:

> *The importance of catabolism of this enzyme pathway...*
>
> *Most people with this disease have bone disease. What is not known is if...*
>
> *Substance A may affect heart disease more than drug B.*

The third paragraph tells the reader what you are seeking to accomplish as a result of the previous two (or more) paragraphs. The last sentence of the last paragraph states the hypothesis and what you are expecting to demonstrate in this paper. This sets the readers' expectations for what will be presented and discussed in the rest of the manuscript.

The last sentence of the third paragraph may start with a sentence like one of the examples below:

> *This investigation analyzed...*
>
> *Data from a large cohort was used to determine the ...*
>
> *This study tested the hypothesis that...*
>
> *This study describes...*
>
> *This study was designed to compare...*

All you need are three sentences per paragraph.

Here is an example of a clear and easy to understand introduction:

> Substance X is a chemical that stimulates the production of Z from TO2 cells. The secretion of Z has been implicated as an important step in the control of cell growth. In certain medical conditions, the secretion of Z is not stimulated and cell growth is not controlled. Unregulated cell growth results in the overgrowth of cells which has been shown to develop into cancer.
>
> In other cell types, substance Z has been shown to be regulated by a variety of intracellular mechanisms including protein kinase C, JAK, CAT, and intracellular calcium. The mechanism of activation of substance Z from TO2 cells is not known. If the pathway for activation of substance Z from TO2 cells can be identified, drugs can be developed to stimulate the secretion of substance Z and control cell growth.
>
> Previous studies in TO2 cells have shown that substance Z is not activated through substance P, JAK, or CAT. Our preliminary evidence indicates activation of substance Z through protein kinase A. The present study was conducted to determine if protein kinase A and C are involved in the production of substance Z from TO2 cells.

When you are satisfied with your introduction, read it out loud to see how it sounds and flows.

The best way to understand the ingredients for a clear and easy to understand introduction is to read many introductions. As you read these introductions, use the *Guidelines for Writing Introductions* to assess the introduction.

Try the next activity to give yourself practice evaluating and writing parts of introductions!

Activity 4:
Introductions

1. Read the introduction of a published journal article.

2. Evaluate the introduction using the *Guidelines for Writing Introductions.*

3. As you evaluate the introduction, write the positive parts of the introduction on a piece of paper.

4. Next, write the parts of the introduction that do not meet the criteria in the *Guidelines for Writing Introductions.*

5. Rewrite each part of the introduction that did not meet the criteria in the *Guidelines for Writing Introductions.* Alternatively, identify the missing information or extra parts included in this introduction. Make a list of the information to include in the introduction.

Chapter 6:
Writing the Materials and Methods

LEARNING OBJECTIVES

- List the main point of the materials and methods section.

- List three points to include in the materials and methods section.

- List two recommendations for writing the materials and methods section.

> ## Chapter Highlights:
> ## GUIDELINES FOR WRITING MATERIALS AND METHODS
>
> - Use subheadings to separate subsections such as Study Design, Study Population, Drugs, Data Collection, Outcomes, and Statistical Analyses.
> - Provide details of the methods used including data collection, inclusion/exclusion criteria, dates of study, drugs studied, tests performed, outcome measures, and statistical analyses.
> - Provide enough detail for others to evaluate and reproduce your work.
> - Cite previously published methods.

The materials and methods section provides a clear explanation of the methods used to conduct your study. Include enough information for an independent person to understand what you did so they can go into the laboratory to repeat your experiments and replicate your results.

Readers will use this information to evaluate the methods used to conduct the experiments. The use of standard methods will strengthen your results. Cite all methods that have previously been published.

Use subheadings to separate subsections.

The materials and methods are easiest to write using subheadings. Depending on your study and the journal targeted for submission, subheadings can be used to separate sections. Examples of subheadings are:

- Study Design

- Study Population

- Drugs

- Data Collection

- Outcomes

- Statistical Analysis

Read your target journal to see the types of subheadings used for different papers including ones like yours. Use only the subheadings that apply to your study.

Provide details of the methods used.

Remember to add details of all the methods used. The methods are complete when you can answer the following questions:

- Who?

- What?

- When?

- Where?

- How?

- Why?

Review your methods as you set up and conduct your experiments to make sure all necessary details have been included. Add any missing information.

Reread, revise, rewrite, repeat.

When you are satisfied with the materials and methods, read it out loud to see how it sounds and flows. Then give it to a colleague who knows your methods to make sure it is complete and easy to understand. Repeat the process described in Chapter 1 until this section is complete.

Try the next activity to give yourself practice evaluating and writing parts of materials and methods!

Activity 5:
Materials and Methods

1. Read the materials and methods section of a published journal article.

2. Evaluate the materials and methods using the *Guidelines for Writing Materials and Methods.*

3. As you evaluate the materials and methods, write the positive parts of the materials and methods on a piece of paper.

4. Next, write the parts of the materials and methods that do not meet the criteria in the *Guidelines for Writing Materials and Methods.*

5. Imagine yourself going into the laboratory to replicate these findings. Do you have enough information to set up and complete these experiments? If not, make a list of the questions that need answering before you can reproduce this work.

6. Make a list and identify the pieces of information that were missing from the materials and methods section. Attempt to rewrite one part of the materials and methods that did not meet the criteria in the *Guidelines for Writing Materials and Methods.* Alternatively, identify the missing information or extra parts included in the materials and methods. Make a list of the information to include in the materials and methods.

Chapter 7:
Writing the Results

LEARNING OBJECTIVES

- List the main point of the results section.

- Identify five components of a clear and concise results section.

- List three recommendations for writing the results section.

Chapter Highlights:
GUIDELINES FOR WRITING RESULTS

General

- Present the results in a logical order.

- Explain the results in the order presented in the tables and figures.

- Present the results of statistical analyses.

- Include only relevant results, with no interpretation.

- Present data clearly in tables or figures.

- Write one paragraph (three sentences minimum) for each major result.

- Lead the reader through the results with clear explanations of all figures and tables.

- Focus the text on the main points of the table or figure.

- Use subheadings, as appropriate.

- Use consistent wording in text, tables, and figures.

Figures and Figure Legends

- Use figures (line graphs, pie charts, bar charts) to show patterns, trends, or relationships in the data.

- Figures are easy to read.

- Figures and figure legends stand alone, without reading the text.

- Figure legends are a brief title for the figure.

- Place figure legends underneath each figure.

- Explain definitions and abbreviations in figure legends.

- For figures, label each axis with units, and add range and tick marks for easy reading.

- When showing more than one group on a figure, use common

symbols such as open, solid, hatched, and cross-hatched patterns to denote different groups.

- Include a key on the graph identifying all symbols.
- Use colors to denote different groups; journals may charge a fee for color figures.

Tables and Table Titles

- Use tables to compare data or values.
- Tables present many numbers; precise values are important.
- Column headings describe the numbers in each column.
- Column headings include the variable and the unit of measure.
- Include an easy to understand title above each table.
- Tables and table titles are understood without reading the text.
- Use footnotes to explain information relevant to entire table such as *P*-value, statistical test, type of measurement, or define abbreviations.
- Tables supplement the text.

Present the results in a logical order.

The results section reports the results of your study. The results tell a story which ultimately leads to the main conclusion(s) of your study. Often, the story is presented in a different order than the experiments were conducted. This is fine. It is most important that the results and the story are presented in a logical order that make sense.

Before you begin putting your results together, check your target journal for any limitations on the numbers of tables or figures allowed. If your study exceeds this number, you may want to combine figures into several parts (such as a, b, c, d), choose an alternate journal, include some data in supplementary materials, or decide to divide the data into two parts for two papers. Before you put your time and energy into the paper, know the expectations of the target journal by reading the guidelines for authors.

Begin preparing your results by reviewing your tables and figures. Decide which tables and figures present your data most effectively. These are the tables and figures to include in your paper. If there are other tables or figures you would like to include, consider using them as supplementary materials. If you have trouble deciding, ask authors or a colleague for help.

In clinical journals it is customary to have the first table present the characteristics of the patient population (Table 2). Subsequent tables and figures present data from the current study in a logical order.

Next, decide the order of presentation. This is the way you would tell someone the story of your study. This order should flow and make logical sense. Once you have an order, explain the story to authors and colleagues to ensure it is understandable and presented clearly. Listen to all comments received and revise the presentation of the results accordingly.

Table 2. Sample Demographic and Clinical Characteristics

	Control Patients	Treated Patients	P value
Total number of Patients	200	89	
Age (years) Mean (SD)	n=200 61 (20)	n=81 68 (20)	0.09
Gender, n (%) Male Female	n=199 188 (94.5) 11 (5.5)	n=88 24 (27.3) 64 (72.7)	0.04
Age at Diagnosis[a] (years) Mean (SD)	n=83 63 (11)	n=13 62 (8)	0.08
Ethnicity, n (%) Caucasian All Others	n=168 110 (65.5) 58 (34.5)	n=77 15 (19.5) 62 (80.5)	0.01

[a] Patients with no diagnosis date were excluded from the analysis.

Data Presentation

In scientific papers, data are presented as either tables or figures. Many people do not know the best way to present their data. It is perfectly acceptable to look at the data as tables and figures before deciding what is easiest to understand. Below are some recommendations for presenting data in figures and tables:

Figures and Figure Legends

Figures are used to show patterns, trends, or relationships in the data. There are many types of figures. A few of the most common types are line graphs, pie charts, and bar charts.

The figures and figure legends should be understood without reading the text. Under each figure there is a figure legend. The figure legend serves as a title for the figure and explains all definitions. The x and y axis of the figure should be labelled and include units and tick marks for easy reading. Use common symbols on the graph such as open, solid, hatched, and cross-hatched patterns to denote data from different groups. Always include a key on the graph so the reader understands the symbol for each group. Use these same symbols consistently throughout the paper. This will make the paper easier for the reader to read and follow your results.

Tables and Table Titles

Tables are used to compare data or values. By describing the variable in each column, the column heading tells you what the number in each column represents. The labels on column headings should be consistent with the language used in each table of the manuscript as well as in the text and figures. The column heading should include the unit of measure such as g, kg, mL, L.

For tables, the title is placed above the table. The title should be clear and easy to understand. Footnotes provide additional information necessary to understand the data such as statistical test, P value, explanation of an abbreviation or a term.

Writing the Results

Once you have your tables and figures and have chosen the order in which they will be presented, you are ready to begin writing the results section. Each table or figure should be explained in one paragraph, which consists of a minimum of three sentences. The paragraph should focus only on the results, with no interpretation. The interpretation is found in the discussion section.

Example of results paragraph for clinical and demographic results (Table 2)

A total of 289 patients enrolled in this study: 200 patients met the criteria for the control group and 89 patients were in the treated group (Table 2). The treated group had more females (72.7%) than the control group (5.5%), and similar age at diagnosis. In the treated group, most patients were non-Caucasian (approximately 81%) compared to the control group, which was 65.5% Caucasian (P=0.01).

If data are presented in a table or figure, the written explanation in the text should focus on the significant main points in the table or figure. Do not present all the data in the text. Only repeat the main point(s) in the written text and refer the reader to the table or figure for the detailed presentation of data. Explain the data in the written text of the results section clearly.

Write the results section by imagining you are taking the reader by the hand and explaining the results to them. For this reason, start each paragraph with an explanation of what each result shows. This is particularly helpful to readers outside the main field of your paper. Examples of sentence starters are below:

Dose-dependent increases in …

To examine if…

The ability of…

To determine if…

The clinical characteristics were…

The cellular response of…

Example of first sentence

To determine if substance A was involved in the activation of enzyme B, the effect of A on enzyme B activity was investigated in the presence and absence of the enzyme B inhibitor, XXX.

Connect all paragraphs so the results are presented in a logical order. Sometimes using subheadings makes it easier to explain the data. It is permissible to use subheadings for each major finding. The title of subheadings may be used to guide the reader, as shown in the following examples:

Examples of subheadings

Results at Baseline

Pre-Treatment Results

Post-Treatment Results

Follow-Up Studies

Try the next activity to give yourself practice evaluating and writing results!

Activity 6a:
Results

1. Prepare a figure from the data below:

Time (hours)	Enzyme activity (mg/µl) Group A	Enzyme activity (mg/µl) Group B
0.5	0.6	2.5
1	0.7	4.6
2	0.4	6.8
5	0.3	1.2

2. Plot the data in several ways. Decide how the data should be represented. [Hint: When possible, order the data from the smallest to largest values.]

3. Write a results paragraph describing this figure and a figure legend.

4. Challenge:

   a. Complete the table by adding a table title.

   b. Write a results paragraph describing the data presented in the table.

Activity 6b:
Results

1. Read the results section of a published article.

2. Choose a figure or table from the results section you have read.

3. Evaluate the results paragraph that describes this figure or table using the *Guidelines for Writing Results.*

4. As you evaluate the results paragraph, write the positive information included on a piece of paper.

5. Next, revise the results paragraph so the rewritten paragraph meets the criteria in the *Guidelines for Writing Results.*

6. When finished, compare your paragraph to the results in the paper.

7. Alternatively, identify the missing information or extra information included in the results paragraph. Make a list of the information to include in the results paragraph.

Chapter 8:
Writing the Discussion and Conclusion

LEARNING OBJECTIVES

- List one main point of the discussion section.

- Identify five elements of a clear and concise discussion.

- List three recommendations for writing the discussion and/or conclusion.

Chapter Highlights:
GUIDELINES FOR WRITING THE DISCUSSION AND CONCLUSION

- Keep the discussion section focused on the topic of the manuscript.
- Restate the hypothesis and explain the main finding(s) of the study in the first paragraph.
- Explain the contribution and significance of your study to the current literature.
- Interpret and discuss each result or set of results in the order presented in the results section.
- Write one discussion paragraph for each major result.
- Each result paragraph places the data in the context of the study and literature.
- Cite relevant references and primary sources.
- Consider all evidence, those that support and refute your data.
- Present alternate explanations for the results.
- Include a limitations paragraph, if appropriate.
- Weave the limitations into the discussion; do not leave it as the last paragraph before the conclusion.
- The conclusion is the last paragraph of the discussion.
- Summarize the results and state the take away message of the paper in the conclusion.
- Draw conclusions from the data presented.
- Discuss your most important conclusion first.
- Explain how your results support the conclusions.
- Include next steps of research, if appropriate.

The goals of the discussion section are to explain and interpret your results as well as indicate the significance of the findings. You will use the current scientific literature to put your findings into context for the reader.

This chapter will give you the structure and tools to write a focused discussion. Think of the discussion as the section in which you take the reader by the hand and walk them through the story of your study. You will start with your hypothesis and significance of your findings, then explain and interpret each result and how it fits into the current literature before wrapping up your discussion. The discussion section allows you to provide more explanations than the introduction, but it still needs to focus the reader on the main points of your study. The discussion needs to be focused to leave the reader with the full understanding of your study, its significance, and the meaning to the field.

The first paragraph of the discussion restates the hypothesis and explains the significance of the findings. This tells the reader exactly what new information was found in this manuscript. This is your sound bite and the most important point and take away message of the paper.

Start the first sentence of the first paragraph with a topic sentence that tells the reader why you did the study. This is a restatement of your original hypothesis, which joins the introduction with the discussion section. Then add your findings and state why they are important.

An example of a first paragraph starter is below:

This study was conducted to determine....Then describe your primary findings and their importance...This is important because...or This is the first demonstration of....

Other paragraphs in the discussion section discuss each major result. Each major result is the data presented in each table or figure or written about in the results section. Each of these results should have a paragraph written about it. Interpret each result and draw conclusions from either one result or group of results. Each paragraph should contain appropriate literature citations. The literature included should support or refute the data obtained and explain why the results of this study are similar or different from previous studies. The discussion should put your results into the context of the current scientific literature.

Example of result paragraph in the discussion

In the experiments reported here, tail temperatures of cold stressed rats decreased within 24 hours of exposure to 4C and returned to control levels within 24 hours after removal from the cold. This elevated skin temperature has been observed previously (cite reference). It is worth mentioning that animals exposed to cold over a period of 5 days did not exhibit this severe drop in skin temperature. This drop in skin temperature may be caused by changes in the temperature regulating system, which may be mediated by corticosterone through the hypothalamic-pituitary-adrenal axis.

The **limitations paragraph** is an important paragraph for clinical studies. It accurately describes what the limitations of the study are and how these limitations may affect the results or interpretation of the study. Some limitations are type of study or small sample size.

Every study has limitations that should be stated for proper interpretation of the results. This paragraph should be placed somewhere in the discussion section, preferably in the middle of the section. It should not be the last paragraph of the section, if possible.

One example of a limitations paragraph sentence starter is:

The limitations of this study are...

Example of limitations paragraph

The use of data from observational research has some limitations, which are common to these studies. Data from all participants are not submitted at the same time points. The data entered are retrospective and patients are not randomized to treatment. Other confounding factors include comorbid conditions, smoking, and alcohol use.

The last paragraph of the discussion is the conclusion. The conclusion summarizes the results and restates the main findings of the paper. If people read nothing else but the last paragraph of the paper, they should be able to understand the significance of this study and know what you did and why it is important. Some people read the last paragraph to see if they want to read the rest of the paper. For this reason, it must contain the take away message of the paper.

Start the last sentence of the last paragraph with a topic sentence that tells the reader what you found. Examples of last paragraph sentence starters are below:

We conclude...
In summary...

Example of last paragraph

In summary, we demonstrated that substance A stimulates the intracellular messenger, B by a mechanism dependent on JAK-A. The evidence includes activation of JAK-A by substance A and complete inhibition by the JAK-A inhibitor, X. The ability to completely reverse the response with the JAK-A inhibitor demonstrates substance A works through the JAK-A intracellular pathway.

Reread, revise, rewrite, repeat.

When you are satisfied with your discussion section, read it out loud to see how it sounds and flows. Then give it to other authors or a colleague along with your results section for review. Continue the process described in Chapter 1 until the section is complete.

The best way to understand the ingredients for a clear and easy to understand discussion is to read many discussion sections. As you read these discussions, use the *Guidelines for Writing the Discussion and Conclusion* to evaluate the discussion section.

Try the next activity to give yourself practice evaluating and writing parts of the discussion section!

Activity 7:
Discussion

1. Read the discussion section of a published journal article.

2. Evaluate one paragraph of the discussion using the *Guidelines for Writing the Discussion and Conclusion*.

3. As you evaluate the discussion paragraph, write the positive parts of the paragraph on a piece of paper.

4. Next, rewrite the parts of the discussion paragraph that do not meet the criteria in the *Guidelines for Writing the Discussion and Conclusion*. Make sure the rewritten paragraph meets the criteria in the *Guidelines for Writing the Discussion and Conclusion*.

5. Alternatively, identify the missing information or extra information included in the discussion paragraph. Make a list of the information to include in the discussion paragraph.

6. Get more practice by evaluating each paragraph of the discussion and rewriting each part that did not meet the criteria in the *Guidelines for Writing the Discussion and Conclusion*.

7. When finished, compare your paragraph to the discussion in the paper.

Chapter 9:
Writing the Abstract

LEARNING OBJECTIVES

- Identify the sections of a formatted abstract.

- List three recommendations for writing the abstract.

- Identify three components of a clear and concise abstract.

Chapter Highlights:
GUIDELINES FOR WRITING ABSTRACTS

- Write the abstract last.

- Abstract is concise.

- Standard abbreviations are defined and used.

- No errors in spelling, punctuation, grammar, word choice, capitalization, and word order.

- Organize the abstract into the following sections:

  Background: (not always included)

  Objective: States the hypothesis clearly

  Materials and Methods: Describes the main features of the study design and the methods used

  Results: Briefly and clearly states the most important results in the same order as in the text of the manuscript

  Discussion: Explains the the relevance and importance of the findings

  Conclusion(s): States the take home message(s) of the paper that are specific and relevant to the results in the paper. If there is room, you can speculate on the significance of your findings

- Observe journal guidelines on word limit, typically 250-350 words.

Write the abstract last.

I write the abstract last after I have written all the other sections of the paper. I do this because the goal of the abstract is to summarize each section of the paper. Once you have written all the other sections of the paper, and all authors agree on the data presented and the flow of the paper, it is easier to pull out the main points of each section for inclusion in the abstract.

Think of the abstract as a miniature version of your paper. If the reader reads nothing else but the abstract of your paper (as some will), the reader will understand the hypothesis, significance of the study, the methods used, main results, and conclusion(s) of your paper, including your main point(s) and take-home message(s). This is important because the abstract will be used as a "calling card," "elevator speech," or "business card" of your work in the scientific literature and the internet. In other words, the abstract will become the public face of your work and many more people will see the abstract than read your paper. This is one way you may become known by other scientists and the public.

A well-written abstract is short, and many journals have a word limit of 250 to 350 words. I always use the formatted abstract (Table 3). This format ensures you will include all the necessary information in the abstract which can be adjusted to reflect the requirements of each journal. If a journal does not accept the formatted abstract, just delete the section headings and make sure the abstract flows in a logical order. Either way, this format will help you write the abstract.

Table 3. Formatted Abstract

Background:	Optional, puts study in context for reader
Objective:	States the hypothesis
Materials and Methods:	Describes how the study was conducted
Results:	Reports most important results
Conclusion(s):	States the take home message of the paper

Formatted Abstract

A description of the information to be included in each section of the formatted abstract is below:

Background: The background section should describe the relevant background of the work. It should be brief, one or two sentences in length. This section is considered optional and may be deleted if the word count for the abstract is exceeded.

Objective: The objective describes the purpose of the study and states the hypothesis of the paper. Some examples of sentence starters for the objective section are below:

To demonstrate that...
To determine that....
The purpose of this paper is...
We hypothesized that...

Materials and Methods: The materials and methods section describes the key important methods used and includes information like study design, cell lines or growth conditions, number of patients or animals in the study, gender of patients or animals, the names and doses of drugs or other chemicals used in the study, the main techniques used, and statistical methods used to analyze the data. The goal of the materials and methods section is to give the reader enough information to evaluate and, potentially, replicate your work.

Results: The results describe the key finding(s) of the paper. These are the most important findings of the paper. Present the tables and figures in the logical order that allows the story of your research to unfold. Start the results section with one of the sentence starter examples below:

> *We found that…*
> *The results demonstrate…*
> *The results show…*
> *The data indicate…*

Discussion: The discussion explains your key results, shows how they fit into the current scientific literature, and tells the reader why they are important. Discuss the results in the same order as presented in the results section.

Conclusion(s): The conclusion is the take home message of the study. This is the most important point of your study which is supported by your data. It is the one point you want someone to remember. If the word count is not exceeded, you can speculate on the significance of your findings in the last sentence of the conclusion.

Start the conclusion with one of the sentence starter examples below:

> *In conclusion…*
> *To summarize…*

When you finish your abstract, read it out loud to hear how it sounds and flows, and revise as needed.

The best way to understand the ingredients for a clear and easy to understand abstract is to read many abstracts. As you read these abstracts, use the *Guidelines for Writing Abstracts* to assess each abstract.

Examples of Formatted Abstracts:

SCIENTIFIC ABSTRACT

Background: Viral vectors like adenoviruses are the most efficient tool for transferring genes in mammalian cells. The xyz cell line has been the substrate of choice for virus production. Due to safety concerns a new cell line for adenovirus production is needed.

Objective: The purpose of this work was to evaluate the ability of a newly developed human cell line to produce adenovirus.

Materials and Methods: D59 cells were grown in suspension cultures in serum free medium to high cell densities (density of 10^6 cells/ml) for a period of one week. Adenovirus production was measured using XXX. Each experiment was conducted in triplicate. Average adenovirus production was evaluated, mean and standard deviations were calculated.

Results: D59 was grown in suspension culture and serum free medium to cell densities of 9 x 10^6 cells/ml. The mean adenovirus production was 3 x 10^6 cells/ml. On average, more than 900 infectious particles (958.32 ± 20.24, mean ± SD) were observed per cell.

Discussion: The results with D59 are promising with respect to cell growth parameters and vector yield during production.

Conclusion: It is possible to develop a bioprocess platform for adenovirus production using a safer human cell line.

CASE REPORT

Background: Multiple myeloma with plasmablastic morphology of tumor cells and extramedullary presentation of disease often have an aggressive clinical course and resistance to chemotherapy. New treatments are needed for patients presenting with this disease.

Objective: We report the case of an elderly patient treated with one chemotherapy agent for treatment of extramedullary IGA-gamma-secreting multiple myeloma with plasmablastic features.

Materials and Methods: Clinical parameters including x, y, and z were assessed in an 86-year-old man who presented with plasmablastic variety of multiple myeloma. The best treatment option was 12-weekly chemotherapy treatments with drug A at a dose of X ug/ml. Clinical responses and adverse events were recorded at each weekly visit, beginning with the initial pre-treatment visit.

Results: After 12 weekly injections of drug A, the elderly patient showed reductions in all clinical parameters including.... No adverse events were reported.

Discussion: This is the first description of successful treatment with drug A of an elderly patient with plasmablastic variety of multiple myeloma.

Conclusion: The use of drug A may be considered as an option for treatment of the plasmablastic variety of multiple myeloma.

Try the next activity to give yourself practice evaluating and writing abstracts!

Activity 8:
Abstracts

1. Read the abstract of a published journal article.

2. Evaluate the abstract using the *Guidelines for Writing Abstracts*. Note if the abstract is a formatted abstract or all one paragraph.

3. As you evaluate the abstract, write the positive parts of the abstract on a piece of paper.

4. Next, write the parts of the abstract that do not meet the criteria in the *Guidelines for Writing Abstracts*.

5. Revise the abstract as a formatted abstract. Include each part of the abstract that did not meet the criteria in the *Guidelines for Writing Abstracts*. Try to identify what is missing in each part of the abstract and correct it. Alternatively, identify what additional information is needed, if any, in each section of the abstract.

Chapter 10:
Selecting a Title

LEARNING OBJECTIVES

- List two uses of a title.

- Identify two components of a title.

- List three recommendations for writing titles.

> ## Chapter Highlights:
> ## GUIDELINES FOR WRITING TITLES
>
> - Write a clear and concise title.
> - Convey the main point(s) of the manuscript in the title.
> - Write the title as a sentence with a subject and verb, if possible.
> - Begin the title with an important word.
> - Use punctuation, as appropriate.
> - Use only standard abbreviations, if needed.
> - Follow journal guidelines and examples in target journal.

The title emphasizes the main point(s) of the paper. Choosing a title carefully is important because the title is used for indexing and searching for your article in the scientific literature. Readers will review the title to decide if they want to read your journal article. Employers or future employers may view the title to see if this area of research is of interest to them. In this way, the title will be associated with your name and you can use the title as a "calling card" or "business card." In other words, the title may be the public face of your work and many more people will see the title than read your paper. This is one way you may become known by other scientists and the public.

The title should be clear and concise and accurately describe your manuscript. This is often achieved by writing the title as a complete sentence with a subject and verb. This serves to convey the main point of the manuscript and help people remember your take home message. If possible, try to use an important word as the first word of the title.

Using abbreviations in the title can make the title hard to understand, especially if the reader is not familiar with the abbreviation.

If an abbreviation must be used in the title, use only common abbreviations from your field and the journal audience. This is the reason many journals recommend no abbreviations in titles. In addition, abbreviations may make it hard to index your paper properly.

To begin writing titles, read many titles in your target journal. Remember, good writing begins with reading. Reading the table of contents will familiarize yourself with the style of titles used by your target journal as well as journals in your field. All these factors will help you write your own titles.

Try the next activity to give yourself practice evaluating and writing titles!

Activity 9:
Titles

1. Read the table of contents of journals in your field. Pay special attention to the titles in your target journal(s).

2. Choose one title to evaluate using the *Guidelines for Writing Titles*.

3. As you evaluate the title, write the positive parts of the title on a piece of paper.

4. Next, write the parts of the title that do not meet the criteria in the *Guidelines for Writing Titles*.

5. Revise the title. Include each part of the title that did not meet the criteria in the *Guidelines for Writing Titles*. Try to identify what is missing in the title and correct it. Alternatively, identify what additional information is needed, if any, in the title.

6. Pick the title of another journal article and revise as needed until you feel comfortable writing titles.

Chapter 11:
Citing References

LEARNING OBJECTIVES

- Identify the role of references in a manuscript.

- Identify two types of references cited and describe when they are used.

Chapter Highlights:
GUIDELINES FOR CITING REFERENCES

- Cite primary articles published in peer-reviewed journals.
- Cite published abstracts and review articles, as needed.
- Fact check all citations before submitting your manuscript.
- Follow journal guidelines for in-text citations and reference list.
- Cite websites as per journal guidelines.
- Use a reference manager to format references.

References are used in the introduction and discussion sections to explain the context of your work within the scientific literature. This is accomplished with a combination of older and current references. The older references may be "classic" references in your field. The newer references identify the current scientific literature and explain what is known and not known in your specific area of research. In the introduction, references are used to identify what data are missing from the literature and explain the rationale for your study. In the discussion, references are used to explain your data in the setting of the current literature. This will help you draw conclusions about your own work.

References can be cited in the materials and methods section. If methods used in your manuscript are based on methods that have been previously published, cite the published article. This gives the reader the context for the methods used and indicates these are standard methods used in the field. Both increase the credibility of the methods used and the data presented in your paper.

Primary articles are journal articles that contain original data on a particular topic. These are the journal articles that have demonstrated a concept, fact, or mechanism. These citations show the data you are explaining have been studied before. You can use these journal articles to indicate the new information your study has added to this field of research.

Review articles are articles that review a body of literature on a specific topic. These articles are good sources to use for an overview of a subject area. They are also a good way to familiarize yourself with current research on a topic. The reference list will contain the most relevant journal articles for this area. However, when specific studies are referred to in your manuscript, the original studies should be read and cited.

Abstracts submitted to conferences may be published in journals that can be cited. This work is usually submitted before the manuscript is written or the study is completed. In addition, abstracts do not provide all the methods and results presented in the published journal article. For these reasons, abstracts are considered preliminary work and should be cited sparingly, if at all.

Fact check all references to verify the facts you are citing are accurate *before* submitting your manuscript. Sometimes authors or colleagues who review manuscripts ask for clarification of a point. In making the suggested change, the meaning of the sentence may be altered, and the reference cited may no longer be accurate. This can happen with small changes in wording. It is easily corrected by taking the time to "fact check" your references and make sure each reference is cited accurately.

Format the references in your manuscript according to journal guidelines. For ease, reference managers, such as EndNote, can be used to make sure the references are cited in the correct format for your target journal. Following all the journal guidelines will increase the likelihood of a positive review.

Scientific journals are numerous. Try to publish your manuscript in the most well-known and respected peer-reviewed journal in your field. Peer-reviewed journals send all the manuscripts published in the journal to a scientist or clinician like yourself for comment or review. Each manuscript is typically reviewed by three reviewers. The reviewers' comments help the journal editor decide how the journal will respond to your manuscript.

Chapter 12:

Now that the Manuscript is Written-
Revise

LEARNING OBJECTIVES

- List three steps in the process for revising your
  manuscript.

- List five recommendations for revising your manuscript.

Chapter Highlights:
GUIDELINES FOR REVISING MANUSCRIPTS

Paragraphs

- Minimum length for a paragraph is three sentences.
- Divide long paragraphs at logical points.
- Connect paragraphs by asking yourself if the closing sentence of one paragraph and the first sentence of the next paragraph link together clearly.
- The progression of each paragraph should be clear and easy to follow.

Sentences

- Begin sentences with words that introduce the main point of the sentence.
- Split long complicated sentences into shorter and easier to understand sentences.
- Correct any run-on sentences.
- Connect sentences for better comprehension; revise sentences as needed.
- Check each verb tense.
- Delete any unnecessary sentences or phrases.

Words

- Check word choice to ensure the chosen words convey the desired meaning.
- Delete all unnecessary words.
- Check all spelling.

Now that you have a draft of your manuscript, it's time to revise. Use the same process outlined in Chapter 1:

- Work for a designated amount of time.

- Put your writing aside to let it "incubate."

- Review it the next day, or at least 24 hours later.

- Reread what you wrote.

- Read it out loud, so you can hear how it sounds and flows.

- Rewrite.

- Revise.

- When finished, give to authors and colleagues to review.

- Consider all comments received.

- Revise as per the comments and suggestions.

- When content with your writing, consider it completed and submit!

Recommendation: Revise your manuscript

When you review your work, ask yourself some questions regarding the facts that are written such as:

- Did you understand what was done?

- Is the study explained clearly?

- Is that what you meant to say?

- Does each paragraph in the manuscript flow in a logical order?

Pay attention to your words, sentences, and paragraphs. Carefully evaluate them with the following questions:

- Can you say it better, using other words?

- Are all the words necessary?

- Are any words misspelled?

- Are there long complicated sentences?

- Can you make any sentences into shorter and easier to understand sentences?

- Are there run-on sentences?

- Can some sentences be joined together for better comprehension?

- Does each verb have the right tense?

- Do sentences begin with words that introduce the main point of the sentence?

- Does each paragraph have a minimum length of three sentences?

- Can long paragraphs be divided into new paragraphs at logical points for easier reading?

- Does the closing sentence of one paragraph and the first sentence of the next paragraph link together clearly?

- Should subheadings be used?

- Can any sentences between paragraphs be revised to make the links clearer and more easily understood?

Answer each of these questions honestly. Remember–it is always better for you to catch your own errors. If the answer to any of the above questions indicates revising is needed, review and revise each section accordingly. The time you spend now will make your publishing experience easier.

Continue reviewing and revising your paper until you have answered all the questions above. Try to be an honest critic. Revising your own work is key to improving your manuscript, your writing, and moving both forward.

When you are happy with your manuscript, give it to a colleague, professor, or friend who has not read the paper before for honest feedback. Listen to the colleague's comments. These comments should be considered your first set of reviewer comments. Revise the manuscript accordingly. When completed, you are ready to submit the manuscript to the journal of your choice!

Chapter 13:

Authorship: Getting Credit for Your Work

LEARNING OBJECTIVES

- Identify the criteria for authorship.

- Define the term "substantial contribution."

> ## Chapter Highlights:
> ### GUIDELINES FOR AUTHORSHIP*
>
> **At the beginning of the project:**
>
> - Review authorship criteria.
>
> - Decide on authorship.
>
> - Determine substantial contribution criteria for each author.
>
> - Write a plan indicating each author's contribution to the manuscript.
>
> **Before manuscript submission:**
>
> - Review the plan of each author's contribution to confirm the contribution and qualification of each author for authorship.
>
> - Revise author list, as needed.

*Adapted from Marusic et al. (2014).

Who Qualifies as an Author?

Authorship gives credit to the people planning, designing, conducting, and analyzing the study. Authors are responsible for the work. The first author is usually the person who did all or most of the experiments or work for the study. The last author is often called the "senior author" because this is the person in whose laboratory the experiments were conducted as well as the person providing funding for the study. The authors whose names appear in between the first and last authors have contributed to the study and their names are often listed in order of their contribution. If authors have contributed equally, the names are listed alphabetically.

Unfortunately, not everyone uses this guideline for authorship. To standardize who qualifies as an author, the International Committee of Medical Journal Editors (ICMJE), World Association of

Medical Editors, Committee on Publication Ethics, and the Office of Research Integrity have established guidelines for authorship. These organizations generally agree on the criteria for authorship developed by ICMJE (www.icmje.org), which state "authorship be based on the following four criteria:

- Substantial contributions to the conception or design of the work; or the acquisition, analysis, or interpretation of data for the work; AND
- Drafting the work or revising it critically for important intellectual content; AND
- Final approval of the version to be published; AND
- Agreement to be accountable for all aspects of the work in ensuring that questions related to the accuracy or integrity of any part of the work are appropriately investigated and resolved."

Substantial contribution is the term that can be confusing and interpreted in many ways. For this reason, the American Medical Association Manual of Style (2007) has defined the term "substantial contribution" as "an important intellectual contribution, without which the work could not have been completed or the manuscript could not have been written and submitted for publication."

Substantial contribution can be confirmed by listing the contribution of each author to the manuscript, as described by the contributorship model (Battisti et al. 2015). This model encourages transparency by listing the contributions of all authors to the study. Examples of contribution categories are:

- Study conception and design

- Conducting or managing study

- Collecting or interpreting data

- Designing and conducting statistical analysis

- Drafting, reviewing, and approving manuscript

- Funding

Another way to confirm authorship is to cover the name of each author. If the name of one author is covered, the contribution of this author will be removed from the manuscript. Repeat this for each author. If a name is covered and the manuscript remains intact, the name of this author belongs in the acknowledgements section. The acknowledgments section includes all contributors to the work who do not qualify as authors.

The authorship criteria reserve authorship for those who deserve credit for the work and take responsibility for it. Most journals require authors to *honestly* state that they have met the ICMJE recommendations for authorship when a journal article is submitted for publication.

Chapter 14:
Responding to Reviewers' Comments

LEARNING OBJECTIVES

- Identify a process for responding to reviewers' comments.

- List three guidelines for responding to reviewers' comments.

Chapter Highlights:
GUIDELINES FOR RESPONDING TO REVIEWERS' COMMENTS

- Answer all comments.
- Write thoughtful and detailed answers to each comment, even if you disagree with the comment and do not want to make the suggested changes.
- Be polite in your response to reviewers' comments.
- Do not ignore the reviewers' comments.
- Make suggested changes to improve the paper.
- List all your changes in an easy to read manner.
- Include the change(s) made and where changes are by indicating page, paragraph, and line number.

You have worked hard to write your manuscript and, of course, you want a positive response from the journal. While a positive response may be received, it is likely to require you, the authors, to respond to comments from the reviewers. Some comments may be reasonable, but you may disagree with other comments. Whatever the comments, you are more likely to have a positive response if you take the time to respond to all the reviewers' comments thoughtfully and politely. This chapter provides guidelines to help you respond to reviewers' comments.

Often, one may feel frustrated or angry by the reviewers' comments. If this happens, put the manuscript aside until you are able to read the comments with an open mind. Remember the goal of reviewers' comments is to produce a better paper. Upon reviewing your paper, the reviewer may find part(s) of the manuscript unclear or open to misinterpretation. Or, perhaps adding another experiment or more data will improve the paper. There are many reasons reviewers add comments. Once you receive comments from reviewers, it is your job to review the comments and answer them to the best of your ability.

Here is a process and guidelines to use to respond to reviewers' comments:

- Number each comment.

- Read each comment carefully.

- Think about each comment.

- Understand each comment.

- Develop answers to each comment.

- Talk to other authors about each comment.

Use this information to develop a table like the one below:

Reviewer Number	Comment number	Reviewer's Comment	What the Reviewer Means	My Response

As you begin to answer the reviewers' comments, think how you would feel if you were the reviewer and you made these comments. You would want the author to know what part(s) of the manuscript were confusing and what information was needed to improve the manuscript. The reviewers are trying to say this to you through their comments. While the comments may not always be said in that way, they are usually made with an eye to improve the manuscript. With that understanding, as you address each comment ask yourself the following questions:

- Does my response answer the reviewer's concern?

- Will the reviewer feel heard?

- Does the reviewer have a point?

  - For example, were the results overstated or was the discussion too long and not focused?

Respond to each comment, whether or not you agree with the comment. Whatever your response, it is helpful for the reviewer to know you read the comment and considered it. Provide a reason or explanation for your disagreement or agreement. This will help the reviewer and editor evaluate your responses and either accept your answers or send you additional comments. Thoughtfully and respectfully respond to all comments.

Review and revise your paper until you have answered all the comments from each reviewer. Ask yourself the questions in Chapter 12 that you used to revise your manuscript.

When you are happy with your response to each reviewer, give the comments, your responses, and the revised manuscript to the other authors, a colleague, professor, or friend for honest feedback. Listen to all comments and continue to revise the manuscript accordingly. When completed, you are ready to submit your responses to the reviewers' comments and the revised manuscript to the journal.

Chapter 15:
Writing the Poster

LEARNING OBJECTIVES

- List the main point of the poster.

- Identify three sections of the poster.

- Identify three components of a clear and concise poster.

Chapter Highlights:
GUIDELINES FOR POSTERS

- Check poster guidelines before beginning; meet all guidelines.
- Limit the amount of information included on the poster.
- Present information in three columns, so it is read from left to right.
- Write a title that is easy to read and understand from a distance.
- Present the main point(s) well, rather than every detail.
- Present information in a logical order that progresses from one section to the next.
- Use headings to mark each section of the poster.
- Examples of headings are Background, Objective, Methods, Results, Discussion, Conclusion, References, and Acknowledgements.
- Use bullet points for easy reading.
- Write an objective that is easy to read and understand.
- Present materials and methods and study design clearly.
- Present results that are easy to follow and understand.
- Use graphics to show data.
- Use the figure legends and table titles to state the main point of each figure or table.
- Use the same scale on all graphs for easy interpretation.
- Present easy to read and understand tables and figures.
- Explain your key results in a few bullet points in the discussion.
- State the take home message of the poster in the conclusion.
- Cite references, as appropriate.
- Acknowledge all who helped and contributed to the study.
- Use font size and style that is easy to read from a distance.

The goal of the poster is to present your work at a scientific meeting. This enables you to participate in and contribute to the meeting by providing fellow scientists with valuable up-to-date information. This chapter provides helpful information for making your poster.

It is important to remember that your poster may be viewed by people alone (without you). Enough information needs to be included on your poster to be understood with or without your presence or explanation. Your poster may be visited by scientists in your field. This will allow you to make contacts and develop your own scientific network of colleagues and peers. Some of these people may be interested in your work and you for future positions.

Your poster is a sample of your work and, as such, is a personal "calling card." Sometimes visitors to your poster provide you with feedback which can give you new ideas to progress your work forward, lead to collaborations, new directions, or applications for your work. The poster is one way you may become known by other scientists and many possibilities can arise from a poster presentation. To present your best self, take the time to prepare a well thought out poster.

Poster Format

Most meetings will have instructions for preparing your poster. This will include the total size of the poster as well as font sizes to use in preparing the poster. Follow these instructions. Prepare the poster to read in three columns from the top left to the bottom left then move to the right for the second and third columns of information. Alternatively, the poster can be prepared in rows reading from left to right and top to bottom.

The poster follows the abstract and generally includes the sections described below. For ease of reading, use bullet points for each section.

Background: The background describes the relevant background of the work.

Objective: The objective describes the purpose of the study.

Materials and Methods: The materials and methods describe the key important methods used so the reader will have enough information to understand what you did and evaluate your work.

Results: The results are the main focus of the poster and are usually the largest section of the poster. The results describe the key important finding(s) of your study. Tables and figures are included and presented in a logical order that allow the reader to understand the story behind your research. Keep the figures and tables as simple and easy to understand as possible. Figure legends and table titles should state the main point of the figure/table and are placed above the figure or table. Figures are easier to read if the same scale is used for all figures. This will avoid any misinterpretation of the data presented on each figure.

Discussion: The discussion explains your key results. Discuss the results in the same order as presented in the results section.

Conclusion(s): The conclusion lists the take home message(s) of your study. These are the most important point(s) of your study that are supported by your data.

References: Cite references used on the poster.

Acknowledgements: Acknowledge all who helped with the study.

Title: The title of your poster may be used by the meeting organizers to develop key words and search terms for the meeting. The title of your poster is the same as your abstract. Both will appear on your resume or curriculum vitae. It is important that the title accurately reflects your work in the area you would like to become known and employed in. In addition, many people at the meeting may use the title to decide if they want to view your poster. For these reasons, a descriptive title with key words that will pique the interest of viewers, fellow scientists, and future (potential) employers is desired. For easy viewing, use a font size that is easy to read and large, such as sans serif, 120 point. This will enable people to view your poster "at a glance."

When you finish your poster, show it to authors and colleagues to review, and revise as needed.

Activity 10:
Posters

1. View one or more posters.

2. Evaluate each poster using the *Guidelines for Posters*.

3. As you evaluate each poster, write the positive parts of the poster on the *Poster Evaluation Form* below.

4. Next, write the parts of the poster that do not meet the criteria in the *Guidelines for Posters*.

5. Revise each part of the poster that did not meet the criteria in the *Guidelines for Posters*. Try to evaluate what is missing in each part of the poster and correct it. Alternatively, identify what parts of the poster could be revised.

Poster Evaluation Form
Poster title:

Question	What You See	Notes to Self
What is the first thing you see on the poster?		
Organization of the poster: easy to understand, challenging (circle one and explain)		
Does the poster have: too many words, enough words, too few words? (circle one and explain)		
How do you read the poster? Record how your eyes view the poster		

Question	What You See	Notes to Self
What makes the poster attractive from a distance?		
Are graphics used? Yes / No (circle one)		
Is color used? too much or too little (circle one and explain)		
What types of graphs were easiest to understand: bar, line, pie, other (circle all that are appropriate and explain)		
What tables were easiest to understand? (Explain)		

Question	What You See	Notes to Self
What tables were hardest to understand? (Explain)		

Chapter 16:
Presenting Your Poster and Yourself

Chapter Highlights:
GUIDELINES FOR PRESENTING POSTERS

- Prepare short, medium, and long explanations of your poster.
- Practice your poster presentation.
- Welcome people to your poster.
- Smile.
- Make eye contact with people walking past your poster.
- Stand next to your poster; don't sit.
- Know your topic completely.
- Practice telling people about yourself.
- Look people in the eye as you speak to them.
- Speak slowly and clearly.
- Make sure visitors to the poster can hear you.
- Point to specific parts on the poster to lead viewers through it.
- Carefully explain your tables and figures.
- Talk to the viewer(s).
- Dress professionally in comfortable clothing and shoes.
- Act professionally.
- Keep the conversation professional.
- Thank the viewer(s) for coming.
- Collect business cards.
- Follow-up with interested people after the meeting.

Presenting Your Poster

Meetings are often busy for attendees. Hence, people may be interested in seeing your poster but may not be able to spend all the time they would like viewing your poster. For this reason, develop three presentations of your poster: short, medium, and long. This will allow you to present the pertinent information to the largest number of people. If people ask questions or want more information on your study, you will be prepared with longer and more detailed explanations of your work.

Your poster presentation uses your poster to explain your work. The short presentation includes the objective, a short explanation of the methods, the main findings, and take-home message(s). The medium explanation includes the background and a discussion of the data. For the long explanation, be prepared to discuss the details of your methods and data and be able to put your data in the context of the current scientific literature. Think about future experiments and be prepared to answer questions about the future of your research and your research interests.

Act interested and excited about your poster presentation by standing next to your poster. Make eye contact with people and smile at them as they walk by your poster. People are more likely to stop, listen to your presentation, and talk to you if they feel welcomed to your poster. After you present your poster, you may have the opportunity to talk about yourself, your interests, and career goals. This is an opportunity to present yourself and network with others in your field.

Presenting Yourself

Prepare for presenting yourself by thinking about the information you want potential employers and colleagues to know about yourself. Ask yourself the following questions:

- What is the most important piece of information for someone to know about me?

- What background information and experiences are important for people to know about me?

- What additional information should people know about my background and experiences?

- What is my current work or profession?

- Why will people be interested in me?

- What are my areas of expertise?

- What are my skills or specialized knowledge?

- What would I like to learn?

- What are my strengths?

- What are my weaknesses?

- What are my professional career goals?

- What do I want to happen after our discussion?

- What are my research interests?

Use this information to develop a short speech about yourself. Use the following as a guide:

I am a student/professional..........with expertise in...
My strengths include...
I have experience in...
My specific accomplishments are...
I have worked for/with...(types of organizations or industries)
I am/would like to...
I am interested in learning more about...
Additional information about yourself (optional)

Practice your speech using the exercises below until you feel comfortable with your presentation. If you are interested in continuing the conversation with the person you are talking to, ask for a business card or contact information so you can connect after the meeting and build your own scientific network.

Activity 11:
Presentation Skills Practice

1. Rehearse your poster presentation and your speech about yourself in front of a mirror.

2. Find a person willing to listen to your speech and give you feedback on the *Criteria for Feedback Form* on the next page.

3. Make your presentation.

4. Listen to feedback from this person.

5. Continue to practice until you are comfortable with your presentation.

Criteria for Feedback Form

Eye contact

Appropriate   Needs improvement

|---|

Language level

Clear   Needs improvement

|---|

Speed of speech

Too slow   Too fast

|---|

Voice volume

Too soft   Too loud

|---|

Speech delivery

Knowledgeable and confident   Unsure

|---|

Appendix 1:
All Writing Guidelines

GUIDELINES FOR WRITING FOR THE SCIENTIFIC LITERATURE

General

- Identify your audience and target journal.
- Write for the readers of your target journal.
- Start by writing the section that is easiest to write.
- Schedule time to work on your writing each day.
- Write one section of the manuscript each day.
- At the end of one writing session, do not look at your writing for at least one day.
- Re-read, edit, and revise what you have written.
- Read your revision out loud to hear how it flows and sounds.
- Repeat this process until you are happy with your writing.
- When revisions are complete, give the manuscript to a colleague or peer to read, review, and provide feedback.
- Repeat process until manuscript is completed.
- Start with an outline.
- Use simple short sentences: subject-verb-object.
- Use complete sentences and paragraphs.
- All paragraphs flow in a logical order.
- Connect sentences and paragraphs with transition words.
- Start a new sentence by repeating words and phrases from the previous sentence.
- Define all abbreviations. Use a maximum of four.
- Check all spelling, punctuation, grammar, word choice, capitalization, and word order.

GUIDELINES FOR WRITING FOR THE SCIENTIFIC LITERATURE

Outlines

- Use a simple outline to organize each section of the manuscript.
- Write each idea as a short bullet point.
- Just write the idea. Don't worry about the English.
- If the ideas have citations, include the references.

Introduction

- Keep introduction short: three paragraphs (nine sentences minimum).
- Explain the disease or area of research and introduce the topic briefly in the first paragraph.
- Start the first sentence of the first paragraph with a topic sentence.
- Introduce the area of research that is not known and explain the significance of this missing piece of information in the second paragraph.
- Clearly state the gap in knowledge, which is the topic of your manuscript
- State the hypothesis in the third paragraph.
- Include a statement of the purpose and significance of the manuscript.
- All three paragraphs flow in a logical order that is easy to understand and follow.
- Reference all facts.
- Cite only literature relevant to your topic.
- Define all questionable terms and abbreviations..
- Do not mention results in the introduction.

GUIDELINES FOR WRITING FOR THE SCIENTIFIC LITERATURE

Materials and Methods

- Use subheadings to separate subsections such as Study Design, Study Population, Drugs, Data Collection, Outcomes, and Statistical Analyses.
- Provide details of the methods used including data collection, inclusion/exclusion criteria, dates of study, drugs studied, tests performed, outcome measures, and statistical analyses.
- Provide enough detail for others to evaluate and reproduce your work.
- Cite previously published methods.

Results

- Present the results in a logical order.
- Explain the results in the order presented in the tables and figures.
- Present the results of statistical analyses.
- Include only relevant results, with no interpretation.
- Present data clearly in tables or figures.
- Write one paragraph (three sentences minimum) for each major result.
- Lead the reader through the results with clear explanations of all figures and tables.
- Focus the text on the main points of the table or figure.
- Use subheadings, as appropriate.
- Use consistent wording in text, tables, and figures.

GUIDELINES FOR WRITING FOR THE SCIENTIFIC LITERATURE

Figures and Figure Legends

- Use figures (line graphs, pie charts, bar charts) to show patterns, trends, or relationships in the data.

- Figures are easy to read.

- Figures and figure legends stand alone, without reading the text.

- Figure legends are a brief title for the figure.

- Place figure legends underneath each figure.

- Explain definitions and abbreviations in figure legends.

- For figures, label each axis with units, and add range and tick marks for easy reading.

- When showing more than one group on a figure, use common symbols such as open, solid, hatched, and cross-hatched patterns to denote different groups.

- Include a key on the graph identifying all symbols.

- Use colors to denote different groups; journals may charge a fee for color figures.

Tables and Table Titles

- Use tables to compare data or values

- Tables present many numbers; precise values are important.

- Column headings describe the numbers in each column.

- Column headings include the variable and the unit of measure.

- Include an easy to understand title above each table.

- Tables and table titles are understood without reading the text.

GUIDELINES FOR WRITING FOR THE SCIENTIFIC LITERATURE

- Use footnotes to explain information relevant to entire table such as *P*-value, statistical test, type of measurement, or define abbreviations.
- Tables supplement the text.

Discussion and Conclusion

- Keep the discussion section focused on the topic of the manuscript.
- Restate the hypothesis and explain the main finding(s) of the study in the first paragraph.
- Explain the contribution and significance of your study to the current literature.
- Interpret and discuss each result or set of results in the order presented in the results section.
- Write one discussion paragraph for each major result.
- Each result paragraph places the data in the context of the study and literature.
- Cite relevant references and primary sources.
- Consider all evidence, those that support and refute your data.
- Present alternate explanations for the results.
- Include a limitations paragraph, if appropriate.
- Weave the limitations into the discussion; do not leave it as the last paragraph before the conclusion.
- The conclusion is the last paragraph of the discussion.
- Summarize the results and state the take away message of the paper in the conclusion.
- Draw conclusions from the data presented.
- Discuss your most important conclusion first.

GUIDELINES FOR WRITING FOR THE SCIENTIFIC LITERATURE

- Explain how your results support the conclusions.
- Include next steps of research, if appropriate.

Abstract

- Write the abstract last.
- Abstract is concise.
- Standard abbreviations are defined and used.
- Organize the abstract into the following sections:
  - **Background:** (not always included)
  - **Objective:** States the hypothesis clearly
  - **Materials and Methods:** Describes the main methods used in the study
  - **Results:** Reports the most important results in the same order as in the text of the manuscript
  - **Discussion:** Explains the meaning of the findings
  - **Conclusions:** States the take home message of the paper. If there is room, speculate on the significance of your findings
- Observe journal guidelines on word limit, typically 250-350 words.

Titles

- Write a clear and concise title.
- Convey the main point(s) of the manuscript in the title.
- Write the title as a sentence with a subject and verb, if possible.
- Begin the title with an important word.
- Use punctuation, as appropriate.
- Use only standard abbreviations, if needed.

GUIDELINES FOR WRITING FOR THE SCIENTIFIC LITERATURE

- Follow journal guidelines and examples in target journal.

Citing References

- Cite primary articles published in peer-reviewed journals.
- Cite published abstracts and review articles, as needed.
- Fact check all citations before submitting your manuscript.
- Follow journal guidelines for in-text citations and reference list.
- Cite websites as per journal guidelines.
- Use a reference manager to format references.

Revising Manuscripts

Paragraphs

- Minimum length for a paragraph is three sentences.
- Divide long paragraphs at logical points.
- Connect paragraphs by asking yourself if the closing sentence of one paragraph and the first sentence of the next paragraph link together clearly.
- The progression of each paragraph should be clear and easy to follow.

Sentences

- Begin sentences with words that introduce the main point of the sentence.
- Split long complicated sentences into shorter and easier to understand sentences.

GUIDELINES FOR WRITING FOR THE SCIENTIFIC LITERATURE

- Correct any run-on sentences.
- Connect sentences for better comprehension; revise sentences as needed.
- Check each verb tense.
- Delete any unnecessary sentences or phrases.

Words

- Check word choice to make the chosen words convey the desired meaning.
- Delete all unnecessary words.
- Check all spelling.

Authorship

At the beginning of the project:

- Review authorship criteria.
- Decide on authorship.
- Determine substantial contribution criteria for each author.
- Write a plan indicating each author's contribution to the manuscript.

Before manuscript submission:

- Review the plan of each author's contribution to confirm the contribution and qualification of each author for authorship.
- Revise author list, as needed.

Appendix 2:
Outline and Manuscript Template

Title Page

Title:

Authors' names and institutional affiliations:

Corresponding author: Include the name, email address and full address, including zip code, of the author who will be dealing with all correspondence and proofs.

Journal: Add any limitations for journal such as word count, number of figures, and maximum number of references.

Abstract (250-350, maximum number of words)

- Write this section last.
- Organize the abstract in the following sections:
  - **Background:** (not always included)
  - **Objective:** States the hypothesis
  - **Materials and Methods:** Describes the main features of the study design. For example, the population studied, male/female, techniques used, drugs or other agents used, statistical methods used to analyze the data
  - **Results:** Reports the most important results
  - **Discussion:** Explains the meaning of the findings
  - **Conclusion:** States the take home message of the paper. If there is room, speculate on the significance of your findings

Word limit usually 250-350 words, depending on journal.

Introduction

Format: Three paragraphs with a minimum of three sentences in each paragraph.

First paragraph–introduces the disease state or topic of research.

First sentence–is a topic sentence...

Examples:

> *This disease is ...*
>
> *Chronic pain is often found ...*
>
> *Complementary and Alternative Medicine is defined as...*

This paragraph provides a concise description of the topic. Sometimes more than one paragraph is needed to describe the topic adequately.

Second paragraph–introduces the gap in knowledge and the literature that your study fills. This paragraph and the next one(s), if needed, explain the significance and importance of filling this gap.

First sentence

Examples:

> *The importance of this is...*
>
> *Most people with chronic pain have ...*
>
> *What is not known is if drug1 may affect this disease more than drug2.*

Third paragraph–tells what you are seeking to do as a result of the previous paragraphs.

Last sentence of last paragraph–states the hypothesis and what you are expecting to find/demonstrate.

Examples:

> *This investigation analyzed...*
> *Data from a large cohort was used to determine the ...*
> *This study tested the hypothesis that...*
> *This study describes...*
> *This study was designed to compare...*

Methods

- Use subheading to separate subsections such as Study Design, Study Population, Drugs, Data Collection, Outcomes, and Statistical Analyses.

- Provide details of the methods including data collection, inclusion/exclusion criteria, dates of study, drugs studied, tests performed, outcome measures, and statistical analyses.

Results

- Present the data in a logical order.

- Use subheadings to state the result of each finding and lead the reader through the results.

- Include only results (no interpretation).

- Present data clearly in tables or figures.

- Write one paragraph for each major result.

- Include only relevant results.

Discussion

First paragraph–restates the hypothesis and explains the significance of the finding. This is the new information found in this paper. This is the most important point and take away message from the paper. The first sentence is a topic sentence.

First paragraph starter

> *This study found that.... Then describe your primary findings and their importance... This is important because...*

Last paragraph–summarizes the results and restates the conclusions. If I read nothing else but the last paragraph of the paper, I should be able to understand the significance of this study and know what you did and why it is important. Some people read the last paragraph to see if they want to read the rest of the paper.

Last paragraph sentence starter

> *We conclude...*
> *In summary...*

Other paragraphs–Each major result should have a paragraph written about it with appropriate literature citations. The literature is used to support the data obtained or explain why the results of this study may be different from previous studies. Interpret each result and draw conclusions from either one result or a group of results.

Limitations paragraph–describes accurately what the limitations of the study are and how this may affect the results or interpretation of the study. Some limitations are type of study and small sample size.

Limitations paragraph sentence starter

> *The limitations of this study are...*

Conclusion

The data summarized in this report demonstrate that ...

Acknowledgements

Acknowledge all who helped and contributed to the study.

Tables (On separate page)

Figure Legends (On separate page)

References (On separate page)

Supplemental information (optional)

To Learn More

1. American Medical Association Manual of Style: A Guide for Authors and Editors, 10th edition, Oxford: Oxford University Press, 2007. doi: 10.1093/jama/9780195176339.001.0001. Published online 2009: www.amamanualofstyle.com. Accessed July 12, 2018.

2. American Medical Writers Association. Essays for Biomedical Communicators: Volume 1 of Selected AMWA Workshops, Witte FM, Taylor ND eds, 2001.

3. American Medical Writers Association. www.amwa.org. Accessed July 12, 2018.

4. Battisti WP, Wager E, Baltzer L, Bridges D, Cairns A, Carswell CI, Citrome L, Gurr JA, Mooney LA, Moore BJ, Pena T, Saner-Miller CH, Veitch K, Woolley KL, Yarker YE. Good Publication Practice for Communicating Company-Sponsored Medical Research: GPP3. *Ann Intern Med.* 2015;163(6):461-464.

5. Byrne DW. Publishing Your Medical Research Paper. Williams and Wilkins Publishers, 1998.

6. Chipperfield L, Citrome L, Clark J, David FS, Enck R, Evangelista M, Gonzalez J, Groves T, Magrann J, Mansi B, Miller C, Mooney LA, Murphy, Shelton J, Walson PD, Weigel A. Authors' Submission Toolkit: A practical guide to getting your research published. *Curr Med Res Opin.* 2010;26 2010;26:1968-1982.

7. Clinical Chemistry Guide to Scientific Writing. https://www. aacc.org/publications/clinical-chemistry/clinical-chemistry%C2%A0guide-to-scientific-writing. Accessed July 12, 2018.

8. Committee on Publication Ethics. www.publicationethics. org. Accessed July 12, 2018.

9. Concept mapping. https://www.brainpop.com/english/writing/conceptmapping/. Accessed July 12, 2018.

10. Health Care Communications Group. Writing, Speaking and Communication Skills for Health Professionals, Yale University Press, 2001.

11. Huth EJ. How to Write and Publish Papers in the Medical Sciences. Williams and Wilkins Publishers, 1990.

12. International Committee of Medical Journal Editors. http://www.icmje.org/. Accessed July 12, 2018.

13. Marusic A, Boznjak L, Jeroncic A. A systematic review of research on the meaning, ethics and practices of authorship across scholarly disciplines. *PLoS One* 2011;6(9): e23477. doi:10.1371/journal.pone.0023477. Accessed July 12, 2018.

14. Murphy R. English Grammar in Use. Cambridge University Press, 2005.

15. Office of Research Integrity. https://ori.hhs.gov. Accessed July 12, 2018.

16. Seals DR, Tanaka H. Manuscript Peer Review: A Helpful Checklist for Students and Novice Referees. *Adv Physiol Educ* 2000;22:52-58.

17. Swan M. Practical English Usage. Oxford University Press, 1995.

18. Tufte E. The visual display of quantitative information. 2nd edition. Cheshire (CT): The Graphics Press, 2001.

19. Walker E, Elsworth S. Grammar Practice for Intermediate Students. Longman Group, England, 1989.

20. World Association of Medical Editors. www.wame.org. Accessed July 12, 2018.

Made in United States
North Haven, CT
03 June 2022

19774816R00076